S. M. HAROON

Truth in Plane Sight

Discover the Top 10 Flat Earth Proofs Revealing Earth Is Not a Globe

First published by Domecraft Publications, Inc. 2024

Copyright © 2024 by S. M. Haroon

All rights reserved. No part of this publication may be reproduced, stored or transmitted in any form or by any means, electronic, mechanical, photocopying, recording, scanning, or otherwise without written permission from the publisher. It is illegal to copy this book, post it to a website, or distribute it by any other means without permission.

S. M. Haroon asserts the moral right to be identified as the author of this work.

S. M. Haroon has no responsibility for the persistence or accuracy of URLs for external or third-party Internet Websites referred to in this publication and does not guarantee that any content on such Websites is, or will remain, accurate or appropriate.

Unless otherwise noted: all Bible verses are quoted from the King James Version (KJV), and all photos are sourced from Pexels, a website offering free, high-quality images.

First edition

This book was professionally typeset on Reedsy. Find out more at reedsy.com

For Anil, mera aashiq

*In loving memory of
Blossom and Caleb*

When the whole world is running towards a cliff, he who is running in the opposite direction appears to have lost his mind.

— C. S. Lewis

Contents

	Acknowledgments	ii
1	Introduction	1
2	Proof 1: Water is Always Level	10
3	Proof 2: The Horizon Always Rises to Eye Level	17
4	Proof 3: The Local Sun	22
5	Proof 4: The Sky is a Perfect Clock	32
6	Proof 5: Understanding Gleason's Map	39
7	Proof 6: Time Zones Reveal the Truth	46
8	Proof 7: The Earth is Motionless	52
9	Proof 8: Selenelion Eclipses – A Globe Impossibility	57
10	Proof 9: More than Half of the Earth Covered in Light	60
11	Proof 10: The Bible is a Flat Earth Book	63
12	Conclusion	71
13	Appendix: Verses for Further Study	74
14	References	78

Acknowledgments

Thank you to the photographers whose works are featured in this book:

Cover photo:
Irene Lasus

Others:
Matan Levanon
Joni Tuohimaa
Osmany Mederos
Jean Paul Montanaro
Max Mishin
Selim Özgün
Rhys Abel
Iana Pugachova
Maria
SHAHBAZ ZAMAN
Lucas Allmann

Thank you also to @BretFeBibleJesus for the use of his work.

1

Introduction

Welcome to My Home

You've picked up this flat earth book, and you're probably wondering who would be crazy enough to believe that the earth is flat in the twenty-first century. Maybe you picked up this book to debunk the supposed proofs that have "flerfs" duped, or maybe you're curious to learn for yourself because you've never looked into it, and before dismissing it, you'd like to see the evidence first to decide for yourself.

Well, before we get to the evidence, I want to invite you to my home. Think of this book as an afternoon coffee (or tea if you prefer) where I'll share my story with you and then my top ten favourite proofs. As you settle into your comfy chair with your warm drink, I'd like to share that I am a Christian, and I'll be sharing my thoughts from a Christian perspective. So, picture this conversation as being me and two other people, one a Christian and the other not. You can be whichever one best fits you. If you're a Christian too, great! If not, that's OK too. Either way, I'm glad to have you.

The reason why I mention my faith is because, whether we know it or not, we each have presuppositions that guide our thinking. These are the beliefs we hold to be true at rock bottom that we know to be true simply because they're true with no other reason needed beyond that. Depending on your beliefs or worldview, you may hold different presuppositions than I do, but we all have them. So, rather than pretend that there's neutrality (which is a myth), it's better to share upfront where I'm coming from, so you don't have to wonder later.

When it comes to Christianity, I belong to a branch called Reformed Christianity, which stems back to the Reformation. In this tradition, we teach a set of theological mottos known as the five *solas*. It's a fascinating topic on its own, but I only want to focus on one of the five: *Sola Scriptura*. *Sola Scriptura*, at its core, means that Scripture alone (aka the Bible) is the ultimate authority for faith and life. It's not to say we dismiss tradition or teaching from the church, but rather, we view them through the lens of Scripture. If something doesn't align with God's Word, it's set aside. Scripture is sufficient—it's the very foundation we stand on as we seek truth and live out our faith.

So, the teaching of *Sola Scriptura* applies to all of life, meaning that whatever subject or aspect of life we're considering, what Scripture says is the ultimate truth that we trust to be true always. Whatever life or experience would seek to teach us, the knowledge we can gain from these areas must first conform to Scripture. This further means that anything that contradicts Scripture must necessarily be false if we have understood the teaching of Scriptures correctly.

With that being said, you don't need to be a Christian to learn from this book. My aim, firstly, is to acknowledge that Scripture has the highest authority when it comes to learning truth, but once we see that Scripture is God's revelation, being fully truthful, we can actually use the knowledge of Scripture as a framework for looking at everything else in life accurately to discern what is true and false. You can think of Scripture as being like a pair of glasses that I'm wearing that brings everything into right focus, but the objects we'll mostly be looking at with those glasses are the evidences that give us clues about what is true. You may not be wearing the same pair of glasses, and that's OK. Look at the evidence for yourself, and look at the evidence as presented here. See what makes sense, and if you see that something is true, follow the Truth wherever it leads.

My Story

As for me, when I was growing up, I was always the planet kid. I made my own encyclopedia of all the planets just for fun, not for school. I wanted to do my grade three research project on Mars, but my teacher said it was too difficult of a subject—I didn't get why. I even wanted to be an astronaut, but my mom said that was too dangerous. I've always loved looking at the skies since I saw the Milky Way as a child, and I loved the years when

I lived in Saskatchewan, Canada, known as the land of the living skies—where the earth is indisputably flat! I've always taken comfort in the natural world because it leads me to a sense of awe, but never once did I question the doctrines of NASA. They knew the truth, and I never thought anything different.

Life has led me down various paths, and especially since 2020, I've started to do my own research when it comes to debatable topics. Ever so briefly, I had once looked into flat earth, but when I found a few videos dismissing it on YouTube, I felt relieved and didn't look any further for quite some time. However, it didn't quite scratch the itch, because I knew in my heart that the videos dismissing the position only mocked it without steelmanning the viewpoint. I knew that if you want to defeat a view, you first have to know it, and I didn't actually know what the evidence was.

Later on, I looked into it again, wanting this time to find the case presented from a flat earther directly, and the evidence is really very simple. It's so simple, in fact, that a child can easily understand it and know the truth. I wasn't ready to have my whole world (literally!) turned right-side-up when I looked into it the first time, but when I looked a second time, I couldn't unsee what I had seen. I felt sick at first, but then greatly comforted, then increasingly amazed time and time again. I felt sick because no one can accept flat earth without also accepting that we have been lied to very greatly, but comforted because I didn't have to feel anxious about always spinning and never feeling it, and amazed because I could see all the more clearly that this world truly is God's creation and his glory is on full display if we only have the eyes to see it. It may be that the greatest obstacle a person has to accepting flat earth is one's own pride. It's difficult to admit that one could be wrong about something so obvious, yet the truth is the truth. So, follow the Truth wherever it leads you because it will make you wise and will teach you the right way to go in life no matter what life throws at you.

Brief Overview of the Model

This book is not going to be an in-depth book on all the scientific details of the competing heliocentric and geocentric models of the cosmos. This is a beginner's book. My goal is to show you enough evidence that, if you have the eyes to see it, you can accept the truth of flat earth for yourself and then do your own research from there. There are many people doing excellent work in this regard, and hopefully many more to come. However, before diving into the evidence, it's necessary to give a layman's description of the model for clarity.

It's also necessary to clarify that not everyone who holds to flat earth is a Christian. There are many lost people in the world, some of whom know that the earth is flat and not much else. I say this lovingly, but it's true. There are some flat earthers who will tell you there may be lands beyond Antarctica and that there may even be aliens living out there, too! Hogwash! Unfortunately, there is good reason for flat earthers to have a bad reputation, so let me be clear: the model I am presenting is based on a biblical view of the created world because, as mentioned earlier, we are working with the principle of *Sola Scriptura*, and Scripture teaches there is only one sun and moon, not more. So, outer lands are impossible. Now, I'm getting ahead of myself.

Here is the basic working model: the earth is flat—and by "earth," I mean all land, not the globe. Please note: the flat earth is not a pancake orbiting the sun in an otherwise heliocentric universe, but rather we are working with a geocentric flat earth model where the earth is at the centre with the other pieces moving around it. The Bible talks about the land being set on pillars (Job 9:6, Psalm 75:3, 1 Samuel 2:8). I don't know whether that part is literal or figurative, but it certainly means the land is stable, not spinning.

Above the earth, there is a strong impenetrable dome called the firmament. This part is more clear in older Bible translations like the King James Version (KJV) in verses like Genesis 1:6–8 and Psalm 19:1–5, whereas modern versions will obscure this detail so as not to conflict with the religion of the globe. The firmament is the reason why gases don't escape our atmosphere and the reason thunder echoes. The stars are in the firmament, and they are not giant balls of burning gases. I don't know what they are exactly, but a phenomenon called sonoluminescence may be a good alternative explanation, which is the process where tiny bubbles in a liquid emit flashes of light when they are rapidly collapsed by sound waves.

If stars are in fact created via sonoluminescence, this would be possible because above the firmament there are waters, as the Bible teaches (Genesis 1:6–7). Importantly, the waters of all the oceans do not run off the flat earth because Antarctica is not at the South Pole (because there isn't one), but rather it forms a ring around the southernmost reaches of earth. It contains the waters because its elevation is very high, such that the waters do not spill over. There is, however, still a North Pole, which is at the very centre of everywhere on earth. If a visual would be helpful, Gleason's map, the same one pilots use to fly, is a very good representation of the flat earth map (see Proof 5). Finally, the sun and moon are local, and actually the same size to each other, as they appear, and the heavens (literally and/or spiritually) are to be understood as being somewhere above the dome and waters above (e.g., Acts 7:49).

This is the basic model we are working with, and I hope these details can serve as a framework within which the proofs will begin to make sense for you. This is no longer the dominant view in Christianity or anywhere else, but this is actually the model that the writers of the Bible had in mind when they wrote Scripture. Now, I know it's a lot to take in. It may seem unbelievable, and you may want to put the book down right now thinking that your credibility is at stake just for having contemplated such a thing. Nevertheless, I encourage you to follow it through all the way to the end and see what you think from there.

A Note on Selection of Proofs

In formal debate, the burden of proof is a rule that requires the person making a positive claim to provide evidence or justification for their assertion. For this reason, I have specifically chosen proofs for this book that present a positive case for the model as described above rather than proofs that would merely disprove the integrity of the heliocentric model. The proofs will primarily be evidence from the physical world. They may present evidence that directly pertains to the earth being a flat plane, or they may be proofs that pertain to other elements of the model, which, if true, necessitate that the world be flat, thereby demonstrating the model's validity and that the earth is flat indirectly while still making a positive case from direct evidence. Finally, I will devote one proof to biblical texts to show that the geocentric model aligns with Scripture, with some commentary on interpretation.

With this approach in mind, I will spend very little time discussing evidence of fraud related to NASA or the promotion of the heliocentric model in general. The fact of the matter is that if the positive evidence is persuasive enough to convince someone that the geocentric model is correct, then by necessity the heliocentric model must be false, and those that promote it must knowingly or unknowingly be incorrect. Certainly, there is plenty of damning evidence against NASA that could be explored another time, but let the reader do one's own research on this matter. I believe the positive evidence alone is sufficiently compelling to convince one against the integrity of NASA, so I will let it speak for itself.

However, I will say at last, it's interesting to note that the word *nasa* in the Hebrew language means to lead astray, to delude, to seduce (as used in Genesis 3:13 when the woman says the serpent deceived her). You may say that's mere coincidence, but I don't believe in coincidence. Admittedly, St. Paul wrote this for another context, but I believe this verse applies well here too when he says, "But I fear, lest by any means, as the serpent beguiled Eve through his subtilty, so your minds should be corrupted from the simplicity that is in Christ" (2 Corinthians 11:3). So, I invite you to think soberly on this matter, for my hope is for you to know the truth that you may not be deceived by anyone.

2

Proof 1: Water is Always Level

When considering the shape of the earth, one of the most compelling proofs that we live on a flat plane is found in the natural behaviour of water. As we observe in everyday life, water consistently conforms to a specific principle: it always finds its level. Whether it's in a glass, a lake, or the vastness of the ocean, water's surface remains flat and undisturbed by the kind of curvature we're often told exists on the globe model.

Water Always Fills Its Container

Let's start with a basic observation: water takes the shape of its container, and once it comes to rest, it settles at a flat, horizontal surface. This isn't merely a phenomenon of small-scale observation; it's a fundamental property of water everywhere. From a cup to a pool to the oceans, the surface of water behaves predictably and reliably. When gravity supposedly pulls everything towards the centre of a globe, water should curve along that surface. However, observations of large bodies of water show no such curve.

Mirror Lakes and the Salt Flats

One of the most peaceful and visually stunning examples of this principle can be seen in mirror lakes. When a lake is completely still, the water acts like a mirror, reflecting its surroundings with almost perfect clarity. The surface is so smooth that the horizon reflects as a straight line, completely level from one side to the other. This reflection would be impossible if the lake's surface were curved, as we would expect to see some distortion in the reflection. The Salar de Uyuni salt flats in Bolivia, spanning about 3,900 square miles (over 10,000 square kilometers), provide a perfect example of this on a grand scale. When these salt flats are covered with a thin layer of water, they become a massive natural mirror. The reflection is so pristine that you can see a perfect, undisturbed image of the sky across the horizon. If there were any curvature on the earth's surface, we would undoubtedly see it in the form of a distorted reflection, yet the reflection remains clear and straight, once again showing the flatness of the land and water.

Sunlight on Water

Another practical observation occurs when the sun's light reflects on the ocean. As the sun moves across the sky, we often see a straight path of light stretching from the horizon to our vantage point. If the earth were curved, however, the light wouldn't follow this straight path. Instead, it would only shine in one concentrated spot, much like how light reflects on smaller curved surfaces, like a ball. On a globe, we wouldn't see light spreading across the water in a straight line as it does but rather in a single area, reflecting the curve of the earth. The fact that we consistently observe a long, straight reflection on the surface of the water further reinforces that the earth's surface is flat and undisturbed by any curvature.

TRUTH IN PLANE SIGHT

Curvature at Different Angles

One of the arguments often made by those who defend the globe model is that ships disappear over the horizon as they move away from us. If this were the result of the earth's curvature, shouldn't we also be able to see the same curvature at a perpendicular angle? Imagine standing on the shore and watching a ship sail away until it supposedly goes over the horizon. When you look left or right at the same distance, you should see that same amount of curvature across the horizontal distance. Yet, the horizon remains flat and level all around you. This is because the disappearance of the ship isn't due to curvature—it's due to perspective. Even more compelling, powerful zoom cameras, like the Nikon P1000, have been able to zoom in on the horizon and bring ships back into view that have supposedly gone over the curve, demonstrating that the ship had only disappeared from sight due to perspective limits, not because of the earth's curvature.

Laser Tests Proving Flatness of Level Water

In addition to the visual observations, laser tests have been conducted to further prove the earth's flatness. One such test was done at Kootenay Lake in British Columbia, Canada, where laser light was projected just above water level at a distance of 28.6 miles (46 kilometres). According to the globe model, the curvature of the earth should have obstructed the laser, preventing it from reaching the target at such a distance.

However, the laser light was clearly visible, providing direct evidence that the surface of the water is flat over great distances. Experiments like these irrefutably prove that the earth is flat, as the curvature claimed by the globe model simply does not exist.

A Common Experience

Water is something we all encounter in our daily lives, whether we're filling a glass, watching rivers flow, or standing in awe of the ocean. Its behaviour is consistent and reliable. Water fills its container, seeks its level, and remains flat over great distances. These observable properties of water strongly suggest that the earth is flat. If we were truly living on a spinning ball without any protective barrier, the water would fly away immediately, much like how droplets fly away from a soaked tennis ball when it's spun. Despite being told that we live on a spinning globe, the natural behaviour of water contradicts that claim. In its simplicity, water provides us with a clear and undeniable picture of the true nature of our world.

3

Proof 2: The Horizon Always Rises to Eye Level

One of the most observable and intuitive evidences for a flat earth is the behaviour of the horizon. Whether you're standing at sea level or high up in an airplane, the horizon always rises to meet your eye level. This phenomenon contradicts what we would expect to see if we lived on a globe, where the horizon would gradually drop away as you gain elevation, forcing you to look down to see it. The horizon, however, behaves in a way that perfectly aligns with a flat plane.

Perspective and the Horizon

To understand this more clearly, consider how perspective works. Perspective dictates that as objects move farther away from you, they appear smaller and eventually disappear from sight. However, even as they shrink into the distance, they remain level with your line of sight. This is exactly what we observe with the horizon. Regardless of how high we ascend—whether on a hill, in a tall building, or even in a plane—the horizon continues to rise to eye level, maintaining its position at the vanishing point in your field of view.

On a globe, the horizon would begin to fall away as soon as you gain altitude, and you would need to look downward to see it. The consistent behaviour of the horizon suggests that the surface of the earth is not curved but indeed level.

The Shrinking Sun

A related phenomenon is the way the sun appears to shrink as it moves toward the horizon. If we lived on a globe, the sun would appear to remain the same size, setting below the horizon in a sweeping curve as the earth rotates. Instead, the sun seems to shrink in size as it moves farther away from us in the sky, consistent with how objects behave on a flat surface due to perspective. The sun doesn't set below a curved horizon but rather moves farther away and shrinks into the vanishing point, similar to how train tracks seem to converge in the distance.

This shrinking effect is especially noticeable when viewed with time-lapse videos or long-range zoom cameras. The sun maintains its position above the flat earth and moves away rather than dipping below a curved horizon, shrinking as it recedes.

Airplane Flights and the Horizon

The next time you're on an airplane, take a moment to observe the horizon. Even at cruising altitude, the horizon remains at eye level, no matter how high you go. If we lived on a globe, you would need to look downwards to see the horizon as you ascend. Yet, pilots and passengers alike can testify that the horizon continues to rise to meet eye level. This behaviour would be impossible on a globe and is consistent with a flat, extended plane.

A great example of this can be seen in the famous Red Bull jump video, where the man ascends to the edge of the atmosphere before making his jump. The curvature visible in that footage, as in many similar videos, is the result of fisheye lenses, not an actual curved horizon. This becomes clear when you look at the shots from inside the shuttle. Even when he is at his maximum height, the horizon remains at the same level in the window as it did before the ascent, proving that the apparent curvature in external footage is nothing more than lens distortion.

What We See

We don't need to rely on advanced technology or complicated experiments to observe this truth. The horizon is something we all encounter, whether at the beach, on a hill, or in the sky. It consistently rises to meet our eyes, no matter how high we go, and behaves exactly as expected on a flat surface. The idea that the horizon should drop as we ascend simply doesn't match our lived experience.

Simple, everyday observations speak to the reality of a flat earth. Whether it's the horizon that always rises to meet our eyes or the shrinking sun moving away from us, the world behaves as if it's flat. The globe model fails to account for these observations, while the flat earth model provides a simple, clear explanation.

PROOF 2: THE HORIZON ALWAYS RISES TO EYE LEVEL

Here we see the horizon rising to eye level, even at cruising altitude

4

Proof 3: The Local Sun

One of the key observations that supports the flat earth model is the behaviour of the sun. Contrary to the globe model, which claims that the sun is 93 million miles away, evidence points to the sun being much closer and local. Observations of the sun's rays, angles, and light patterns strongly suggest that the sun is a small, local light source, not a distant star in space. This understanding of the sun aligns perfectly with a flat earth model.

Crepuscular Rays as Evidence

One of the most compelling pieces of evidence for a local sun comes from the observation of crepuscular rays. These are the sunbeams we often see when sunlight passes through clouds or gaps in the atmosphere. When observed, these rays clearly radiate outward from a single point, converging down at angles. If the sun were truly 93 million miles away, the rays should appear parallel to one another by the time they reach earth.

However, the fact that we observe crepuscular rays spreading outward from a central point suggests that the sun is much closer and that its light is not coming from a distant source.

Basic geometry supports this observation. By tracing the angles of crepuscular rays back to their origin, we can calculate that the sun is likely only thousands of miles above the earth, not millions of miles away as the globe model suggests. These observations offer simple yet compelling evidence that the sun is local, consistent with the flat earth model.

Seasons and Temperature Variations Explained by a Local Sun

The flat earth model provides a clear and intuitive explanation for both the changing seasons and temperature variations by proposing that the sun moves in circular paths above the earth. Instead of relying on the globe model's complex mechanics of a tilted axis and elliptical orbit, the local sun theory posits that the sun's proximity to different regions of the earth at different times of the year determines seasonal and temperature changes.

During summer in the northern hemisphere, the sun moves in a tighter circular path above the Tropic of Cancer, staying closer to the northern regions for longer periods. This proximity results in warmer temperatures, more daylight hours, and the experience of summer. As the sun's path gradually shifts toward the Tropic of Capricorn, the reverse occurs—northern regions cool down and experience winter, while the southern hemisphere enters summer.

This movement of the sun provides a simple and observable explanation for why temperatures fluctuate throughout the year. Areas closer to the sun's path receive more direct sunlight, resulting in higher temperatures, while areas farther from its path receive less, leading to cooler conditions. This localized movement of the sun, rather than a distant orb and tilted globe, explains both seasonal shifts and daily temperature variations in a way that aligns with what we observe.

The Implausibility of Light-Year Distances Further Supports a Local Sun

Another critical aspect that reinforces the idea of a local sun is the inconsistency we find with light-year distances when comparing the sun and other stars. According to the mainstream view, the sun is approximately 8 light-minutes away from earth, meaning its light takes 8 minutes to reach us. Even with this immense distance, the sun appears relatively small in the sky. Now consider that the next closest star, Proxima Centauri, is supposedly 4.5 light-years away—an incomprehensible distance by comparison. If the sun is 8 light-minutes away and appears so small, how could a star that is 4.5 light-years away be visible at all? By all logic, such a star should be invisible to the naked eye, far too distant to produce any visible light.

The problem is only heightened when we consider Polaris, the North Star, which remains stationary and visible in the sky despite being allegedly 433 light-years away according to the globe model. If the stars were truly such vast distances away, they should be undetectable by the naked eye. Instead, both the sun and stars appear consistent with a much closer, localized system. This issue highlights the fundamental difference between what we observe and what the globe model asserts. It makes far more sense to view the sun and stars as part of a contained system, much closer to earth than mainstream science claims.

Changing Size of the Sun

As noted earlier in Proof 2, the shrinking sun is a key observation in understanding both the flat earth model and the local sun. The sun appears to diminish in size as it moves toward the horizon, which aligns with how objects behave when moving away on a flat plane. On a globe model, the sun should remain the same size throughout the day, given its immense distance. The fact that it shrinks as it moves away suggests it is much closer and localized.

This shrinking effect is not only noticeable in person but becomes even clearer through modern technology. As mentioned previously, time-lapse videos and long-range cameras provide an extended view, capturing how the sun appears to shrink as it moves farther away. These tools offer a more detailed observation than what the naked eye can track, reinforcing the local sun model. They provide clear evidence that the sun's size diminishes with distance—something that would not occur if the sun were truly 93 million miles away.

The Sun in the Firmament

Psalm 19:1-6 offers a vivid description of the sun's placement and movement within the firmament:

"The heavens declare the glory of God;
 and the firmament sheweth his handywork.
Day unto day uttereth speech,
 and night unto night sheweth knowledge.
There is no speech nor language,
 where their voice is not heard.
Their line is gone out through all the earth,
 and their words to the end of the world. In them hath he set a tabernacle* for the sun,
Which is as a bridegroom coming out of his chamber,
 and rejoiceth as a strong man to run a race.
His going forth is from the end of the heaven,
 and his circuit unto the ends of it:
 and there is nothing hid from the heat thereof."

i.e., a tent—referring to the firmament

This passage describes the sun as being in the firmament, running a circuit across the skies, supporting the idea that the sun is a local light source rather than a distant star. Since we will explore the principles for interpreting such texts further in

PROOF 3: THE LOCAL SUN

Proof 10, we'll keep the commentary brief for now.

However, with this biblical depiction in mind, consider three images that clearly show the sun's position relative to the clouds. According to the heliocentric model, the sun should never be positioned behind or within the clouds. Yet, these images reveal just that: clouds can be seen behind the sun, and at times the sun appears nestled within the clouds, strongly indicating its proximity to earth.

If you're still on the fence about a local sun, examine the photos and let them show you the truth with your own eyes. These visual observations further confirm the sun's local nature, reinforcing the flat earth model's understanding of the world.

Here, in picture one, we clearly see clouds both in front and behind the sun.

PROOF 3: THE LOCAL SUN

In picture two, we see the sun literally nestled in the clouds – impossible with the globe model!

Again, in picture three, we see with blissful simplicity light clouds resting behind the sun.

Everyday Evidence

In our everyday experience, we witness the sun's rays, its changing size in the sky, and the temperature variations that come with its movement. These simple, observable facts point to a local sun that is much closer to earth than the globe model claims. The shrinking of the sun as it sets, the directness of crepuscular rays, and the seasonal changes all reinforce the conclusion that the sun is a local object moving above a stationary earth.

PROOF 3: THE LOCAL SUN

The flat earth model offers a clear and coherent explanation for these observations, challenging the widely accepted view of the sun as a distant star. The local sun is yet another piece of everyday evidence that aligns with the flat earth model and reveals the true nature of our world.

5

Proof 4: The Sky is a Perfect Clock

One of the most compelling proofs of a flat earth is found in the sky itself. The stars, moon, and sun all move in predictable patterns, operating like a perfectly synchronized clock. Throughout history, people have relied on the sky to mark the passage of time, seasons, and even their location. These celestial bodies move in ways that align with the flat earth model, where the sky operates above a stationary earth, offering a straightforward explanation for timekeeping and navigation that challenges the globe model.

The Stars: A Yearly Timekeeper

The stars offer an incredible way to measure the passage of time on a yearly scale. Their consistent positions in the night sky serve as a natural calendar, marking the seasons and helping ancient civilizations track the time of year with precision. Constellations like Orion or the Big Dipper appear at predictable times each year, signaling the changes in seasons. This reliability has made the stars an indispensable tool for navigation and timekeeping throughout human history.

In the flat earth model, the stars are fixed within the firmament, rotating above the stationary earth in precise, unchanging patterns. This stability is essential for their use as a yearly timekeeper. If we were on a globe, hurtling through space at high speeds while rotating around the sun, we would expect the stars to shift in position over time. Yet, when we look up at the night sky, we see the same constellations in the same positions, year after year. This consistency is best explained by a flat, stationary earth, where the stars move predictably across the firmament.

Polaris, the North Star, is a particularly compelling example. It remains nearly motionless in the sky, providing a fixed point of reference for navigation. As the earth is supposedly moving in multiple directions through space, the fact that Polaris stays in one place defies the logic of the globe model. On a flat earth, however, this makes perfect sense, as the stars rotate around the North Star in a fixed pattern that repeats every year.

The Moon: A Monthly Timekeeper

The moon offers another example of how the sky acts as a perfect clock. Its phases are consistent and predictable, repeating in a 29.5-day cycle. Each phase of the moon provides a way to track time, not just on a monthly basis but also in relation to the seasons. The moon's consistency, like the stars, suggests a system where the celestial bodies are local and follow precise, repeating paths.

In the flat earth model, the moon moves in a circular path above the earth, much like the sun. Its illumination comes from a source other than the sun, as evidenced by the fact that moonlight is cold, in contrast to the sun's warm light. This can be observed by measuring temperatures in direct moonlight versus in the shade at night, with moonlit areas being cooler. This phenomenon provides further proof that the moon's light is unique and not a mere reflection of the sun's rays. The flat earth view offers a coherent explanation of how the moon's movements and light source serve as a dependable clock in the sky.

The Sun: A Daily Timekeeper

The sun's movement across the sky has long been used to track the time of day. On a flat earth, the sun moves in a circular path above the earth, gradually illuminating different regions as it passes. As it follows its path, the sun provides the most basic and familiar form of timekeeping—the passage of day into night. The flat earth model explains this as the sun moving in a circuit, not as the earth spinning on its axis.

The position of the sun in the sky at any given moment tells us the time of day, just as it has for millennia. People have used sundials and other methods to track time based on the sun's position, and this consistent pattern has always aligned with a flat, stationary earth. The simple, predictable motion of the sun over the flat earth reinforces the idea that the sky above us operates like a clock, governed by a natural order that reflects the flat earth model.

A Coordinated System

What makes the sky such a perfect clock is the coordination between the stars, the moon, and the sun. Each celestial body moves in a precise and predictable pattern, and their movements align perfectly with the flat earth model. On a flat earth, the sky revolves around us, with the stars, sun, and moon all moving in their respective paths. This synchronization has provided humanity with reliable timekeeping for thousands of years, a function that would be far more complex to explain in the globe model, with its spinning earth and chaotic movements through space.

The flat earth model offers a clear, simple explanation for why the sky functions so perfectly as a clock. It eliminates the need for complicated explanations involving vast distances and multiple motions. Instead, it presents a picture of a coordinated system where the celestial bodies move above a stationary earth, providing us with the time and seasons.

The sky's predictable mechanics work together like the gears of a clock

The Sky's Reliable Pattern

The sky functions as a perfect clock, offering reliable timekeeping through the positions of the stars, the phases of the moon, and the movement of the sun. This consistent, predictable system aligns seamlessly with the flat earth model, where the

PROOF 4: THE SKY IS A PERFECT CLOCK

sky operates above a stationary plane. For centuries, people have relied on this natural clock to measure time and navigate the world, and its precision continues to support the reality of a flat earth. The simple, observable movements of the sky provide everyday evidence that the flat earth model aligns with what we see.

The full moon hangs in a starry sky over the flat horizon

6

Proof 5: Understanding Gleason's Map

One of the strongest visual representations of the flat earth model is Gleason's map, a widely accepted flat earth projection. It offers a practical and coherent depiction of the world as we experience it. The map not only aligns with how we navigate the earth but also reflects how distances, flight paths, and oceanic travel work in the real world. Understanding this map helps explain many phenomena that don't fit well with the globe model, making it a crucial tool for visualizing the flat earth.

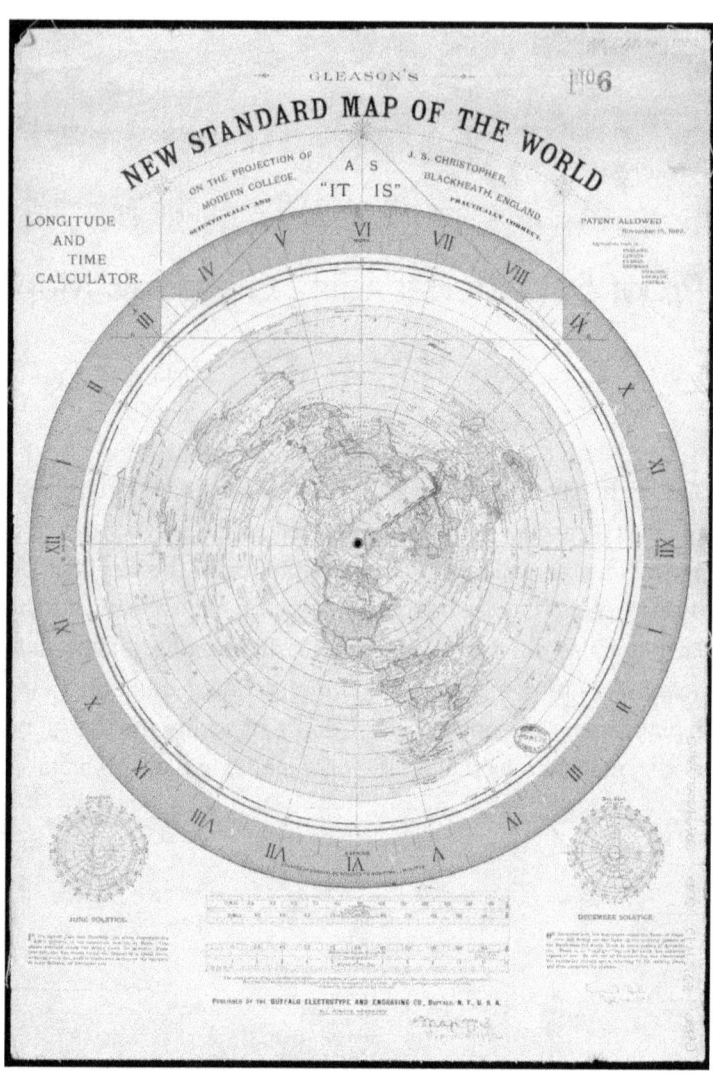

Source: Public domain

Gleason's Map and Real-World Navigation

Gleason's map represents the earth as a flat disc, with the North Pole at the centre and the continents arranged around it. This projection allows for the accurate depiction of distances and travel routes, particularly when it comes to aviation and oceanic navigation. Pilots and ship captains often navigate using maps that resemble this projection, and their routes make much more sense on a flat earth than they would on a globe.

For example, direct flight routes between cities in the southern hemisphere, such as from South America to Australia, seem much longer and more circuitous on the globe model. On Gleason's map, however, these routes become direct and logical. Similarly, emergency flight landings that seem inexplicable on the globe model, such as planes landing far off their expected course, make perfect sense when plotted on Gleason's flat earth map.

Emergency Flight Landings

One of the most well-known examples of unexpected emergency landings comes from the book *16 Emergency Landings Proving Flat Earth* by Eddie Alencar. In one detailed account, China Airlines Flight 68, traveling from Taiwan to Los Angeles, made an emergency landing in Alaska due to a medical emergency onboard. The planned route for this flight should have taken it in a relatively straight path over the Pacific Ocean. On a globe, Alaska appears far to the north of this path, making the detour seem illogical.

However, when viewed on Gleason's flat earth map, the route between Taiwan and Los Angeles passes directly over Alaska, making the emergency stop appear perfectly reasonable. This alignment makes sense in the flat earth model, where the continents are arranged around the North Pole and distances are more direct than they would appear on a globe. In this instance, the globe model struggles to provide a logical explanation for why a plane heading to Los Angeles would divert so far north. Yet, the flat earth map not only explains the emergency stop but shows how it aligns naturally with the flight path.

Another example involves Qantas Flight 64, which was traveling from Johannesburg, South Africa, to Sydney, Australia. This flight made an emergency landing in Perth, despite Perth being significantly off the expected route in the globe model. However, when plotted on Gleason's map, the flight path runs directly over Perth, explaining why this location was a practical place for the plane to land. These kinds of emergency landings, which seem puzzling on a globe, make perfect sense on the flat earth map.

These examples of emergency landings are consistent with the flat earth model's depiction of distances and routes, providing practical, real-world proof that aligns with Gleason's projection.

PROOF 5: UNDERSTANDING GLEASON'S MAP

Underwater Cable Lines and Gleason's Map

Another example that supports Gleason's map comes from the underwater cables that connect the continents. These cables are crucial for international communications and the internet, and their paths often don't follow what we'd expect on a globe. Instead of running in straight lines across the oceans, these cables take paths that appear unnecessarily long and complex on the globe model. However, when plotted on Gleason's map, the routes become much more logical, following direct paths across the flat earth.

For instance, cables between Europe, Africa, and North America follow routes that make sense on Gleason's map, despite looking inefficient on a globe. This again points to the fact that real-world infrastructure aligns with the flat earth model, offering further evidence of the practicality and accuracy of this projection.

A Logical and Consistent Map

Gleason's map is not just a visual aid; it is a practical tool that reflects the reality of how we navigate and travel across the earth. Whether it's aviation routes, emergency landings, or underwater cable lines, the flat earth model provides a more coherent explanation for the world as we experience it. The map aligns with real-world phenomena that the globe model struggles to explain, reinforcing the flat earth perspective.

PROOF 5: UNDERSTANDING GLEASON'S MAP

A Map That Aligns with Reality

The existence and continued use of maps like Gleason's show that the flat earth model is grounded in observable reality. From flight paths to international communication cables, the flat earth map provides a straightforward and accurate representation of the world. Gleason's map aligns with what we see and experience, offering another piece of everyday evidence that supports the flat earth model.

7

Proof 6: Time Zones Reveal the Truth

On a globe, time zones are meant to divide the earth into equal sections based on longitude. As the earth rotates, the sun moves across the sky, and time zones are supposed to change predictably by one hour for every 15 degrees of rotation. To make this system work, the International Date Line was created in the middle of the Pacific Ocean, where a sudden jump of 24 hours occurs when crossing it. This allows the globe model to maintain its 24-hour day.

In the flat earth model, however, this reset wouldn't be necessary. The sun moves in a circular path above the earth, and time zones follow this motion. The regions closer to the center of the flat earth (near the North Pole) would require fewer time zones because the sun's path is shorter. As the sun moves farther outward, the circumference it covers expands, requiring more time zones for the larger areas. The time flows smoothly, without the need for sudden jumps or resets.

PROOF 6: TIME ZONES REVEAL THE TRUTH

Why the Globe Needs a 24-Hour Jump (and Flat Earth Doesn't)

In the globe model, the 24-hour jump at the International Date Line is required because the earth is spherical and rotates on its axis. Without this date line, traveling westward would eventually lead to confusion, where the same location could have two different days. The International Date Line creates an artificial boundary to fix this, ensuring that the time system stays consistent around the globe.

On a flat earth, the sun's movement doesn't require such a drastic reset. As the sun moves in a circular pattern, time changes gradually from region to region. There's no need for a sudden 24-hour jump because you're not traveling around a sphere—time simply flows naturally based on the sun's position over the flat plane.

The International Date Line's Strange "Box"

One of the oddest aspects of the International Date Line is that it doesn't run straight north to south as you might expect. Instead, there's a section where the line juts out eastward, creating a "box" where a couple of time zones are squeezed in. This strange deviation doesn't make sense on a globe model, where time zones are supposed to follow smooth, consistent divisions. It appears that this "box" was created artificially to make the time system work in that specific area.

On a flat earth, there would be no need for such manipulation. The sun's consistent circular motion would create natural time zone divisions, and no artificial adjustments or "boxes" would be necessary.

The Northern Time Zone Compression

In the far northern regions, particularly places like Svalbard, Norway, and northern Alaska, we see an interesting compression of time zones. In the globe model, as you move toward the poles, time zones should become more frequent because the distances covered by each zone get smaller. However, in reality, there are fewer time zones than expected in these northern areas. There are approximately nineteen time zones in the northern regions, far fewer than the expected 24-hour model.

For example, Svalbard shares the same time zone as mainland Norway, even though it's much farther north. Similarly, Alaska, which spans a large area, only operates in two time zones, despite covering a significant portion of the northern hemisphere. This compression makes more sense in a flat earth model, where the sun's path near the North Pole is much smaller, leading to fewer time zone divisions.

PROOF 6: TIME ZONES REVEAL THE TRUTH

Australia's Odd Time Zone Behaviour

Australia, which is roughly the same size as the United States, has far fewer time zones despite its size. While the U.S. has four distinct time zones (Eastern, Central, Mountain, and Pacific), Australia uses just three main time zones. However, even more curiously, South Australia operates on a half-hour difference (ACST) from its neighboring states, rather than a full hour.

This half-hour division is more than just an anomaly; it hints at something deeper. In the southern hemisphere, the sun's path covers a much larger circumference than in the northern regions. There are approximately thirty-two time zones in the southern regions, far more than the 24-hour model can account for. To make this fit into a 24-hour system, smaller, half-hour time zones are used in some places. These half-hour differences are an attempt to squeeze more time zones into the southern region without breaking the globe's 24-hour illusion.

The flat earth model, where the sun's circular movement above the earth covers different regions with larger and smaller circumferences, makes sense of this issue. It explains why Australia has half-hour time zones—because they are trying to force-fit 32 or more time zones into a system designed for just 24.

Rethinking the Nature of Time Zones

From the strange "box" in the International Date Line to the compressed time zones in the north and the half-hour anomalies in Australia, these inconsistencies point to manipulation in the globe model's timekeeping system. The flat earth model, where time follows the sun's smooth circular path, provides a more logical explanation. Without needing artificial resets or forced jumps, time zones on a flat earth would flow naturally based on the sun's position, offering a simpler and more intuitive approach to timekeeping.

PROOF 6: TIME ZONES REVEAL THE TRUTH

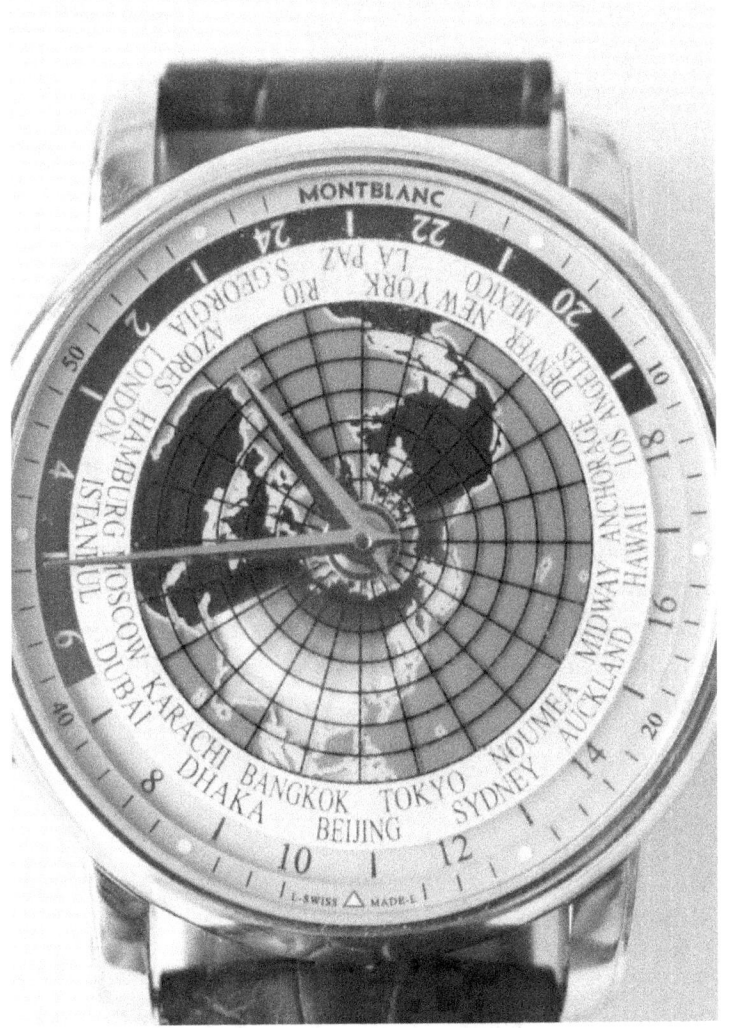

A watch showing time zones on a partial flat earth map

8

Proof 7: The Earth is Motionless

The globe model teaches that the earth is spinning at a dizzying speed of 1,000 miles per hour (1,600 kilometers per hour) at the equator while simultaneously hurtling through space at approximately 67,000 miles per hour (108,000 kilometers per hour) around the sun. However, if the earth is moving so fast, why don't we feel it? And how can it be that everything seems motionless when we stand on the ground?

Michelson-Morley Experiment: No Detectable Movement

One of the most telling pieces of evidence that the earth is motionless comes from the famous Michelson-Morley experiment of 1887. This experiment was designed to detect the movement of the earth through space, as scientists at the time believed that the earth moved through a substance called "the ether." If the earth was moving through space, they expected to detect a shift in the speed of light due to this movement.

However, their experiment revealed no detectable movement of the earth at all. The result was so shocking that it has since been explained away by theoretical constructs like relativity and the bending of space-time to make the math fit the model. The point is this: just because the math works doesn't mean the theory is true. Someone may tell you, "1 fairy + 1 fairy = 2 fairies," but that doesn't mean fairies are real. We shouldn't let the wool be pulled over our eyes just because the math works. The math can be made to fit any narrative, but that doesn't make the narrative true.

The Earth Feels Still Because It Is Still

We don't feel the earth moving because it isn't moving. If you've ever ridden on a high-speed train or in a fast-moving vehicle, you know that even at speeds of hundreds of kilometers per hour, you can still feel the movement. For comparison, the fastest vehicle on earth, the Thrust SSC, holds the land speed record, traveling at 763 miles per hour (1,228 kilometers per hour), which is just above Mach 1. The earth, by comparison, is said to be spinning at Mach 1.3 and orbiting the sun at a staggering Mach 88! If you don't believe you could be on a supersonic vehicle without feeling the motion, why should we believe that we can be on a planet moving through space at tens of thousands of miles per hour and feel nothing at all?

Globe Defenders and Mathematical "Magic"

Globe defenders often argue that the reason we don't feel the earth's motion is because everything moves in unison, creating a smooth experience, similar to how you don't feel the motion of an airplane once you're at cruising altitude. However, this explanation doesn't hold up when you compare the sheer difference in speeds. The fastest commercial airplane flies at around 600 miles per hour (965 kilometers per hour), and yet, when you experience turbulence, the motion is clear. Now compare that to the globe's claim that the earth is spinning at 1,000 miles per hour and orbiting the sun at 67,000 miles per hour—and supposedly, we feel nothing? That kind of speed would cause unimaginable sensations.

Even worse, we are told that the solar system itself is moving at 490,000 miles per hour (790,000 kilometers per hour) through the galaxy. How is it possible to move at such staggering speeds and not detect any motion? The answer is simple: we don't feel it because it's not happening. We are being told to trust the math, to accept theoretical physics and relativity as explanations, but the reality is much simpler—the earth is motionless.

Biblical Evidence: The Earth is at Rest

This idea is not new. The Bible speaks of the stillness of the earth multiple times, reaffirming that the earth does not move. For instance, Zechariah 1:11 says: "And they answered the angel of the LORD that stood among the myrtle trees, and said, We have walked to and fro through the earth, and, behold, all the earth sitteth still, and is at rest."

Some Christians may object to using such verses to support a flat earth position, claiming they are meant figuratively. To them, I would ask, what do your senses tell you? Not only do our senses contradict this explanation, but insisting that such verses are figurative effectively says that some Bible verses are objectively false, even if taken figuratively. Personally, I prefer to stick with the Bible, even if ten thousand PhDs say otherwise.

Don't Be Fooled

The bottom line is that we are constantly told not to trust our senses. We're told that although the earth feels motionless, it's actually moving at unbelievable speeds. However, when our direct experience and experiments like Michelson-Morley tell us otherwise, why should we dismiss them? The earth feels still because it is still. Don't let the complex math blind you from the obvious truth. The simplest explanation is often the correct one—the earth doesn't move.

"[B]ehold, all the earth sitteth still, and is at rest." – Zechariah 1:11

9

Proof 8: Selenelion Eclipses – A Globe Impossibility

A selenelion eclipse occurs when both the sun and the moon are visible in the sky during a lunar eclipse. According to the globe model, this should be impossible. During a total lunar eclipse, the earth is supposedly positioned between the sun and the moon, casting a shadow on the moon. For the sun and moon to be visible at the same time would mean they are not perfectly aligned, contradicting the globe model's claim that the earth is blocking the sunlight.

The Impossibility of the Globe Model

Globe proponents explain this with atmospheric refraction, claiming the earth's atmosphere bends light, allowing both the sun and the moon to be seen simultaneously. However, this explanation defies the limits of known physics. While refraction can slightly bend light, it cannot account for the massive discrepancy needed to allow for this alignment. What we are being asked to believe is that light is bending around the entire curvature of the earth, creating the illusion that the sun and moon are visible together during the eclipse. This is an impossibility.

If the earth were truly blocking the sunlight, both the sun and moon would not be visible in the sky at the same time. The concept of "refraction" is used as a convenient fix, yet it strains the limits of reason.

A Flat Earth Perspective

On a flat earth, the explanation is much simpler and grounded in observable reality. The sun and moon are visible at the same time because they are not aligned in the way the globe model suggests. On a flat earth, both the sun and moon move above the earth in their respective circuits. A selenelion eclipse is simply a visual effect caused by their movements, not a result of one blocking the other or casting shadows over long distances.

What Do Your Senses Tell You?

Think about what you can observe: you can see both the sun and the moon during these eclipses. Your senses are telling you that something is off with the globe explanation. The supposed atmospheric refraction is just a way to patch up the inconsistencies in the heliocentric model. Instead of blindly trusting mathematical constructs that bend reality, we should rely on our senses and the straightforward explanations they offer us.

The reality is simple—what we observe doesn't match what we are being told. The globe model demands explanations that go beyond reason to preserve its narrative, while the flat earth model gives us a much clearer understanding. Don't be fooled by overly complex theories. The sun and moon can be seen together during an eclipse because the earth is flat, and they are moving in paths above us, not due to refraction or shadows bending across impossible distances.

10

Proof 9: More than Half of the Earth Covered in Light

There are certain days of the year when more than 50% of the earth's surface is illuminated by sunlight, a phenomenon that directly challenges the globe model. On the equinoxes, when day and night are supposed to be evenly split, this anomaly occurs, with sunlight covering a greater area than should be possible if the earth were a sphere.

The Globe Model's Inconsistency

According to the globe model, the earth is constantly rotating, with one half of the earth experiencing sunlight and the other half experiencing darkness. This system is meant to keep the balance of light and dark across the planet, especially during equinoxes, when day and night are supposed to be of equal length worldwide. However, on certain days, more than half of the earth is lit by the sun, which should not be possible if the earth were truly a globe. Globe defenders claim that this is due to atmospheric refraction, once again bending light around the curvature of the earth to create the illusion of more sunlight. However, this explanation fails to account for the overwhelming area of light that can sometimes extend over vast distances.

Flat Earth Explanation

The flat earth model offers a simpler and more logical explanation for this phenomenon. On a flat earth, the sun moves in a circular path above the surface, with its light covering large areas, including more than 50% of the earth's surface at certain times of the year. This is due to the way light spreads out over the plane, especially during times like the equinoxes. The idea that atmospheric refraction can cause such a widespread distribution of light on a globe doesn't align with basic observational evidence.

What we observe is clear: more than half of the earth is illuminated by sunlight, something that is far easier to explain on a flat earth model, where light behaves predictably and spreads across large portions of the plane, without requiring complex theories or bending the laws of physics.

Trusting Your Senses

Just as we've seen in other proofs, trusting what you observe is essential. We can clearly see that more than half of the earth is covered in sunlight at certain times of the year. The globe model's explanation of refraction requires us to dismiss this clear observation. Instead of relying on convoluted explanations, the flat earth model provides a more straightforward and consistent understanding of how light behaves.

11

Proof 10: The Bible is a Flat Earth Book

Addressing the Hermeneutics Objection

Many Christians argue that using flat earth to interpret Scripture misapplies hermeneutics, claiming it takes figurative language literally. However, understanding the Bible's cosmology as flat earth is integral to how the biblical authors viewed creation. Holding a heliocentric view introduces contradictions, at times even in Jesus' own words, such as what he speaks of the stars in Matthew 24:29. He says:

"Immediately after the tribulation of those days shall the sun be darkened, and the moon shall not give her light, and the stars shall fall from heaven, and the powers of the heavens shall be shaken."

In a heliocentric model, where stars are enormous and millions of miles away, this passage seems nonsensical. In a flat earth model, however, where stars are closer and smaller, this passage makes perfect sense. For those who attempt to reinterpret "stars" here as meaning comets or meteors, they must admit that this implies Jesus was wrong in saying that stars would fall. If Jesus meant something other than stars, it would undermine His words in this crucial prophecy. The flat earth cosmology, with its understanding of stars as smaller, closer lights, maintains that Jesus' words are both literal and true.

Additionally, the flat earth view requires far fewer assumptions, especially regarding the passage in Joshua 10:12-14, where the sun is commanded to stand still. This is not poetic but historical narrative. There's no need to reinterpret what the passage clearly states. The text tells us that the sun was commanded to stop, not the earth. This is crucial evidence that the biblical cosmology involved a moving sun over a stationary earth, consistent with the flat earth model.

Understanding Poetic Versus Literal Language

Some teachers, such as Mike Winger, claim that passages supporting flat earth are figurative and should not be taken literally. However, biblical poetry is still meant to communicate truth. For example, in Job 38:10, God sets boundaries for the sea with "bars and gates." Since the sea doesn't have literal bars or doors, this text is clearly figurative but conveys the real truth that the sea has divinely established limits. Similarly, flat earthers often cite Job 38:14, where the earth is described as taking shape "like clay under a seal." While this verse uses poetic language, it conveys a literal truth about the earth's structure. Discernment is required to distinguish between figurative language and the truth being communicated, but every passage—whether figurative or literal—communicates something real.

It requires discernment to differentiate between what parts of Scripture are poetic and which parts are literal, yet both always communicate truth. To claim that all passages supporting flat earth are merely figurative and bear no reflection on reality strips these verses of meaning. If the Bible's cosmology is entirely wrong, as some argue, that creates a serious issue for anyone holding to *Sola Scriptura*. The reality is that the biblical authors, under divine inspiration, understood and wrote about a flat earth.

Dismissing all flat earth verses as figurative without recognizing the truth they contain strips them of meaning. Psalm 19:1-5 is a case in point. The psalmist declares that the heavens proclaim God's glory, and the firmament shows His handiwork. Furthermore, the passage says that the sun is in the firmament, running its course like a strong man. The globe model, with its heliocentric assumptions, renders these verses almost meaningless—neither figurative nor literal truth is communicated. The flat earth model, however, preserves the literal meaning of these verses by maintaining the sun's movement within the firmament, as Scripture describes.

If all the verses supporting flat earth cosmology were merely figurative, it would imply that the Bible is fundamentally wrong about the nature of creation. Since the biblical authors, under divine inspiration, believed in a flat earth, those denying flat earth cosmology must confront the implications of rejecting the Bible's teaching. For those holding to *Sola Scriptura*, the question becomes: is the Bible accurate in its teaching on cosmology, or are modern, secular theories overriding biblical truth? By embracing flat earth cosmology, Christians can affirm *Sola Scriptura* without contradictions, while rejecting a worldview rooted in atheistic assumptions. Though it may be difficult to accept this perspective, the pursuit of truth is worth the challenge. With that in mind, let's take a closer look at a key biblical passages that support a flat earth cosmology.

The Firmament and the Heavens

In Genesis 1:6-8, God creates the firmament, a solid structure that separates the waters above from those below. The firmament is described as something tangible and enduring, not a metaphorical concept. Psalm 19:1-5 declares that "the heavens declare the glory of God; and the firmament sheweth his handywork." This passage emphasizes that the sun is placed within the firmament, running its course like a strong man. The flat earth model, with its firmament concept, takes this description literally, viewing the firmament as a real, physical structure separating the waters above from the earth below. The globe model struggles with this concept, as it provides no explanation for the firmament, rendering these verses almost meaningless in both literal and figurative interpretations.

Additionally, Psalm 104:2-3 describes God stretching out the heavens like a curtain and laying the beams of His chambers in the waters. This reinforces the biblical view of the heavens being stretched over the earth, further supporting the flat earth cosmology that sees the heavens as part of a created firmament.

The Earth Set on Pillars

Several verses describe the earth as fixed and immovable, set on pillars. 1 Chronicles 16:30 says, "The world also shall be stable, that it be not moved," and Job 9:6 speaks of God shaking the earth, which is set on pillars. These passages clearly present the earth as a stable, unmoving foundation, consistent with the flat earth model. The globe model, on the other hand, teaches that the earth is constantly spinning and orbiting, which contradicts these biblical descriptions.

Again, Psalm 75:3 says, "The earth and all the inhabitants thereof are dissolved: I bear up the pillars of it." The flat earth model takes this literally, viewing the earth as physically resting on pillars. The immovability of the earth, combined with its placement on pillars, is a strong indicator of the flat earth cosmology described in Scripture.

The Sun and Moon

The sun and moon are described throughout Scripture as moving across the sky, not the earth moving around them. In Joshua 10:12-14, already referenced, Joshua commands the sun to stand still, and God grants this request. This passage is part of a historical narrative and presents a clear understanding that the sun moves, while the earth remains fixed. The flat earth model requires no re-interpretation here—what is written is exactly what is happening.

Moreover, Ecclesiastes 1:5 says, "The sun also ariseth, and the sun goeth down, and hasteth to his place where he arose." This verse describes the sun's daily movement across the sky, supporting the flat earth view that the sun travels over a stationary earth, not the earth rotating around the sun. The globe model's explanation, where the earth's rotation creates the appearance of the sun's movement, doesn't align with the plain meaning of these Scriptures.

Christ Enthroned Above Creation

Christ is the Creator and Sustainer of all things, as stated in Colossians 1:16-17: "For by him were all things created, that are in heaven, and that are in earth, visible and invisible." A flat earth cosmology reflects Christ's dominion over a creation that is consistent with what we observe. By trusting what we can see and experience, we affirm Christ's sovereignty over a world that functions according to His design, not human theories. Christ, having been "seated at his right hand in the heavenly places" (Ephesians 1:20), presently reigns from a real, tangible heavenly realm above the earth. In the flat earth model, the heavens above and the earth below fit into a divine structure created and upheld by Christ's power.

Isaiah 66:1 further supports this, stating: "Thus saith the LORD, The heaven is my throne, and the earth is my footstool." This reinforces the flat earth model's view of the heavens and the earth as two distinct, layered creations—God's throne is above, and the earth is below, as a stable foundation for His creation.

Antarctica as Earth's Boundaries and the Earth as a Circle

Several biblical passages point to the idea of Antarctica being the boundary of the earth, as well as descriptions of the earth being a circle or disk. Job 26:10 states, "He hath compassed the waters with bounds, until the day and night come to an end." This passage suggests that the waters of the earth are confined within boundaries, which aligns with the flat earth model's understanding of Antarctica encircling the earth's perimeter and holding in the waters.

Isaiah 40:22 also says, "It is he that sitteth upon the circle of the earth, and the inhabitants thereof are as grasshoppers." The word "circle" here is significant, as it describes the earth in a flat, circular form, not as a globe. It's also important to note that in Hebrew, there is a word for "ball," which is used elsewhere in Scripture, but Isaiah 40:22 specifically uses the word for "circle," reinforcing the flat earth model. Additionally, the Dead Sea Scrolls give further weight to this view, describing the earth as a "disk," perfectly aligning with the flat earth model.

These verses, along with many others, suggest that the earth is encircled by a limit, namely Antarctica, forming a boundary for the waters, and is described as a flat circle rather than a globe. This interpretation aligns with a flat earth cosmology, which emphasizes the literal and observational realities of the world as described by Scripture.

More Verses for Further Study

Some say that flat earth is a conspiracy theory, yet, if so, it has been said, "No conspiracy causes more to read the Holy Bible." This is true because the scope of the biblical data supporting flat earth cosmology is extensive. Several more passages provide additional insight into the flat earth view, but the scope of this book limits a full exploration. For further study, I encourage you to examine the passages contained in the appendix, which provide a rich foundation for a biblical flat earth worldview.

12

Conclusion

This book is only an introduction. There are many more questions left to explore—about NASA, gravity, the moon, planets, and more. The evidence provided here should give you confidence that solid answers are available as you continue your research. The journey is just beginning, and I hope this book has sparked your curiosity.

Most Christians will readily affirm that the big bang and evolution contradict the Bible's teachings (read the book of Genesis), but few recognize that the heliocentric globe model is also derived from the atheistic worldview. It completes a triad of beliefs designed to undermine biblical truth. Think about it: the globe model teaches the insignificance of the earth, despair at the idea that the earth will eventually die when the sun burns out and anxiety at the supposed spiraling movements of all the systems that make up the universe—no stability, no sense of God's ordered creation. The flat earth model, by contrast, upholds the earth as central and stable, consistent with a worldview that places trust in the Creator's design.

If the proofs and evidence presented in this book have at least made you curious, I encourage you to explore more about flat earth and its implications. The purpose of this book is to introduce this little-known truth. I encourage you to dive deeper into the science, examine NASA's legitimacy for yourself, and follow the rabbit trails as far as you care to go. It is beyond the scope of this book to answer every question. I only exhort you to make every effort to love and follow the Truth with discernment, wherever it leads you. What you do with the truth shared here is now up to you.

Christian apologist Joe Boot once wisely said, "What we don't know is killing us." Questioning our senses leaves us disconnected from the truth that God has revealed. When we are constantly told to dismiss what we can see and experience firsthand, we lose touch with reality. This tactic is not new. George Orwell captured this manipulation perfectly when he said, "The Party told you to reject the evidence of your eyes and ears. It was their final, most essential command." In today's world, we're told the same. Trusting what God has shown us through creation is essential.

Romans 1:20 reminds us:

"For the invisible things of him from the creation of the world are clearly seen, being understood by the things that are made, even his eternal power and Godhead; so that they are without excuse."

God created the world in such a way that we can understand the truth of it. His creation was not designed to deceive us but to reveal Himself. The sun, the moon, and the earth testify to the Creator's design.

Christ sits enthroned in the heavens, and our worship belongs to Him alone. If you have not yet put your faith in Christ, I invite you to come out of the darkness and into the light. Christ, in His great love, died on the cross to save lost sinners. He rose again, offering eternal life to those who believe in Him. See the truth of God's creation and His love for you, and know that through Christ, light and redemption can be yours today.

Support the Search for Truth

If you've found this book helpful, I would greatly appreciate a positive review. Your feedback not only helps others discover the truth about the earth but also supports further inquiry into the nature of God's creation. Thank you for taking the time to share your thoughts, and may your pursuit of truth continue bearing fruit.

13

Appendix: Verses for Further Study

The following verse references were compiled by @BretFeBible-Jesus on Instagram, used with permission:

The Earth Does Not Move:
   1 Chronicles 16:30, Psalm 33:6-9, Psalm 93:1, Psalm 96:10, Psalm 104:5, Psalm 119:89-90, Isaiah 14:7, Isaiah 45:18, Zechariah 1:11

The Earth's Shape Described:
   Job 38:13-14 "clay under a (ring's) seal"
   Isaiah 40:22 "disk of the earth" in the Dead Sea Scrolls

The Firmament Dome's Shape Described:
   Psalm 19:4, Isaiah 40:22

The Earth on Its Pillars:
   Job 9:6, Psalm 75:3, 1 Samuel 2:8

Tides and Winds Caused by Sun and Moon and Cold Four Storehouses,

APPENDIX: VERSES FOR FURTHER STUDY

Vents & Treasuries in the Firmament Dome:
Job 37:9-12, Ecclesiastes 1:5-6, Jeremiah 31:35, Psalm 135:7, Revelation 7:1, Jeremiah 10:13, Jeremiah 51:16

Windows in the Firmament Dome to Pour Out Water, Fire, Hail, Brimstone Coals and Stars Upon the Face of the Whole Earth:
Genesis 7:11, Genesis 8:2, Genesis 19:24, Job 38:8, Isaiah 24:18-19, Revelation 6:13, Luke 21:35, Psalm 18:12-14, Job 38:22-23, Joshua 10:11, 2 Peter 3:5-6, Psalm 78:23

Stretch Heavens Like a Tent Curtain with Its Pillars/Stakes and Spread Earth Like a Potter's Plate and Laid the Flat Foundations of the Earth Like a Cottage's:
Job 9:8, Psalm 104:2-5, Isaiah 40:21-22, Isaiah 42:5, Isaiah 44:24, Isaiah 45:12, Isaiah 51:13, Isaiah 51:16, Jeremiah 10:12, Jeremiah 51:15, Zechariah 12:1, 2 Samuel 22:8, Psalm 18:7, Psalm 18:15, Psalm 82:5, Job 38:4-7, Isaiah 24:18-20 (cottage reference), Isaiah 48:13, Proverbs 8:29, Micah 6:2, Psalm 102:25, Job 26:11

Antarctic Ice Wall and Bounds/Ends of the World:
Job 26:10, Psalm 19:4, Proverbs 8:27, Psalm 74:17, Acts 17:27, Deuteronomy 4:32

The Height of the Giant Trees Seen All Across the World Proves Zero Curvature, Firmament Dome Height and the Width of the Flat Earth and Depth of the Earth:
Job 11:7-9, Job 22:12, Job 38:18, Proverbs 25:3, Psalm 103:11-12, Jeremiah 31:37, Ecclesiastes 7:24, Revelation 9:1, Revelation 20:1, Psalm 102:19, Psalm 148:1, Daniel 4:11-20 (God gives flat earth dreams, not globe dreams), Genesis 11:4 (Tower of Babel

they saw dome during flood—windows opened)

The Firmament Is Solid and God's Handiwork Like a Blacksmith Where God's Throne Rests on Top of It and Is Described as Being Made of Glass, Crystal, and Sapphire:

Exodus 24:10, Job 37:18, Job 38:37, Psalm 19:1-4, Ezekiel 1:22-28, Ezekiel 10:1, Acts 18:3 (God is a firmament dome tentmaker), Amos 9:6 (vaulted dome), Job 22:14 (vaulted dome), Revelation 4:6, Revelation 15:2, Psalm 150:1 (sanctuary), Hebrews 8:2-5 (the tabernacle was patterned from the firmament), NOT Isaiah 22:18 (God is not a ballmaker)

Brightness of the Firmament, "Sun Dogs" and Sun Reflection on Dome:

Daniel 12:3

Every Eye Sees Jesus Under the Whole Firmament/Heaven (Not via Phones):

Revelation 1:7, Deuteronomy 4:19

The Sun, Moon and Stars Move in Circuits/Courses and Each Give Their Own Light and Are Set Under (Sun and Moon) and Above (Stars) the Firmament Dome:

Joshua 10:12-13, Job 31:26, Psalm 19:4-6, Psalm 148:3, Isaiah 38:8, Judges 5:20, Genesis 1:14-17, Job 38:31-33, Isaiah 13:10, Psalm 136:7-9, Job 9:7, Habakkuk 3:11

Ocean Waters Under the Flat Earth:

Genesis 7:11, Genesis 8:2, Exodus 20:4, Job 26:5-14, Psalm 148:7, Proverbs 8:28

APPENDIX: VERSES FOR FURTHER STUDY

Waters Above the Firmament Dome—Not Rain:

Genesis 1:2, Genesis 1:6-7, Psalm 104:2-6, Amos 5:8, Amos 9:6, Prayer of Azariah 1:38, Psalm 107:24-27, Psalm 18:9-11, Psalm 148:4, 2 Peter 3:5-6, Job 28:11, Job 37:10 (flat face of the deep ocean), Job 38:30 (flat face of the deep ocean), Genesis 2:5-6 (no rain before the flood)

14

References

Alencar, E. (2020). *16 emergency landings proving flat earth.*

Boot, J. (2024, May 10). *WCCA spring banquet address* [Speech]. Toronto, ON, Canada, St. Paul's Bloor Street.

@BretFeBibleJesus. (2024, May 9). *The past two days I've been compiling the top Flat Earth & Firmament Dome verses from the Bible into a single meme that fits on IG. It's approximately 116 passages of truth showing you the reality of where you live. I get asked about the best verses all the time so here is something for you to study and learn new FE verses.* [Instagram Post]. Instagram. https://www.instagram.com/p/C6vCRfDudOB/?igsh=NHZraWl3cWRzaGlw

Gleason, A. (1892, November 15). *Gleason's new standard map of the world: on the projection of J. S. Christopher, Modern College, Blackheath, England ; scientifically and practically correct ; as "it is."* [Map]. Digital Commonwealth. Retrieved October 1, 2024, from https://www.digitalcommonwealth.org/search/common

REFERENCES

wealth:7h149v85z

H., B. (2024, September 1). *28.6-mile laser test proves earth is flat* [Video]. https://www.youtube.com/watch?v=BsMyyG0XMgE

King James Bible. (1769). *King James Version*. Cambridge University Press. https://www.kingjamesbibleonline.org/

Lewis, C. S. (n.d.) When the whole world is running towards a cliff.... (Commonly attributed).

Orwell, G. (1949). *1984*. Secker & Warburg.

Whitsitt, A. (2023, May 17). *Schooling globers - episode 11 - time zones*. [Video]. YouTube. https://www.youtube.com/live/0iTmJl1QeAE

Winger, M. (2020, March 10). *Refuting those "flat earth Bible verses": You should have checked the context.* [Video]. YouTube. https://www.youtube.com/watch?v=qePiXK_8Agc

www.ingramcontent.com/pod-product-compliance
Lightning Source LLC
Chambersburg PA
CBHW070350230526
45471CB00006B/2503